Australian Turtles

Their Care in Captivity

AUSTRALIAN TURTLES
Their Care in Captivity

© 2016 C. Egan

First published as 'A Guide To Keeping Australian Tortoises'
(ISBN 0959354204) Gnome Publications, 1978.

Title: Australian turtles : their care in captivity /
C Egan (author and illustrator);
Trish Hart (cover artist).

Edition: 6th edition (revised).
ISBN: 9781925110968 (paperback)
ISBN: 9780645212914 (hardcover)
Notes: Includes bibliographical references.

Subjects: Turtles--Australia.
Turtles as pets--Australia.
Animal welfare--Australia.
Dewey Number: 639.392

LEAVES of GOLD
PRESS

Copyright (C) 2021 Leaves of Gold Press
ABN 67 099 575 078
PO Box 345, Shoreham, 3916, Victoria, Australia

Australian Turtles

Their Care in Captivity

C. Egan

Contents

Introduction

Most people know very little about the care of the native semi-aquatic turtles that are sold throughout Australia as pets. The majority of these animals die when they are very young, mainly due to the ignorance or laziness of their owners. Don't be one of these owners. Your pet will live a long time if you do your part by devoting a little time to its welfare.

As pets, however, turtles may not give you a great deal of enjoyment. They are wild creatures and often appear miserable when kept in captivity. They will try their best to escape and make their way back to their natural environment. It is not recommended that turtles (or tortoises) are kept as pets, but if you do keep them, having taken away their freedom it is only fair that you must ensure their lives are long and healthy.

Keeping turtles properly can be time-consuming. Tortoises and turtles live for well over a hundred years, if they are given the right conditions. Some have been known to live for nearly 200 years. Unlike other pets, a turtle is a life-time commitment.

Think of this when you acquire one.
You may have to remember your turtle in your will.

Turtles are social animals. It is kinder to purchase two rather than one as a pet, as one by itself will be lonely and stressed, just as a human being would be. They enjoy each other's company.

When babies, turtles are about the size of a ten or twenty cent piece.

In three years, a short-necked turtle can grow to up to 165mm (6 1/2 inches) from nose to tail and 10cm (4 inches) across the carapace.

Turtles do not have teeth, but bite their food with a hard horny ridge around their mouths, also using their claws to tear at it.

Aquatic turtles love to be in the water most of the time, but they must be able to come out of the water onto land if they wish. They do not have gills, like fish, but breathe air with their lungs and store it inside their bodies to use while they are under water.

Tortoise, Turtle or Terrapin?

Turtles belong to the class Reptilia, which includes crocodiles, alligators, lizards, snakes, the tuatura of New Zealand and of course tortoises.

"The terms 'turtle' and 'tortoise' are often used interchangeably and can cause some confusion. In the past, all freshwater turtles were called 'tortoises' and marine turtles were called 'turtles'. The more recent convention has been to restrict the term 'tortoise' to the purely land-dwelling species. As such, Australia has no tortoises."

~ Australian Museum website, retrieved 25th April 2016

Marine Turtles have flippers or webbed feet with long claws, and they live in the salty waters of the ocean.

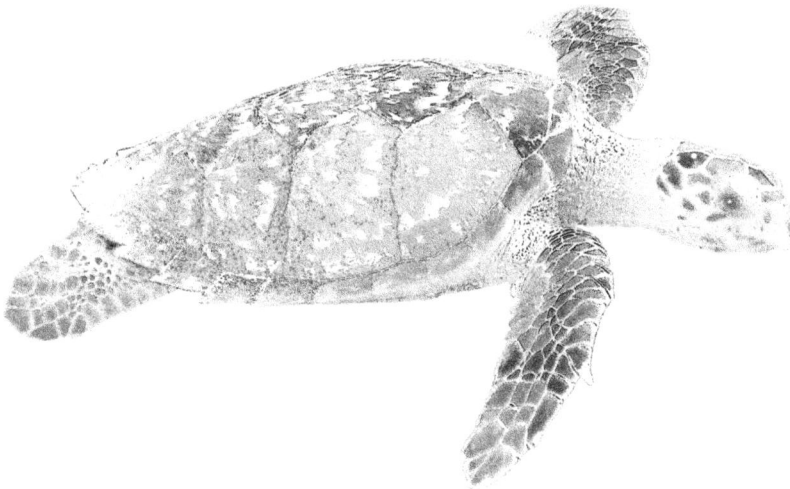

Hawksbill sea turtle (Eretmochelys imbricata)

Semi-Aquatic Turtles do not have flippers; instead they have elephant-like feet with small claws. They live both on land and in fresh water such as rivers, lakes and streams.

Murray River Turtle (Emydura macquarii)

Terrapins are a type of turtle native to South America, and are not available in Australia. It is illegal for members of the Australian public to keep tortoises, turtles or any other reptiles that are not native to Australia.

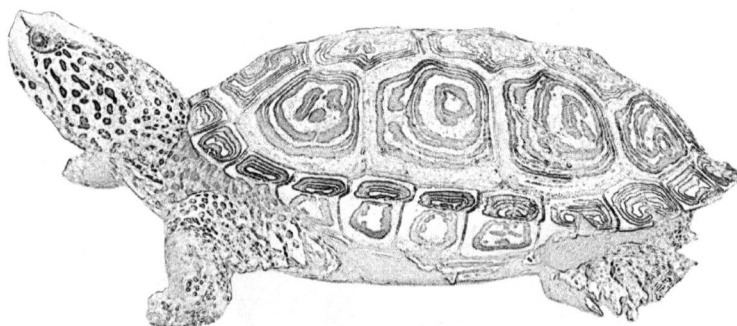

Diamondback Terrapin (Malaclemys terrapin)

Land-Dwelling Tortoises also have clawed feet instead of flippers. Also known as 'terrestrial tortoises', these creatures are usually herbivorous and cannot swim.

African Spurred Tortoise (Geochelone sulcata)

Descriptions

The upper half of a turtle's shell is called the carapace. The underneath half is called the plastron.

There are around 23 native species of semi-aquatic freshwater turtle in Australia. They all belong to a sub-order called Pleurodira. Pleurodires are side-necked turtles; that is, when they withdraw their head for protection, they bend their neck sideways.

Pleurodires are only found in the southern hemisphere. They are a more ancient form of turtle than the Cryptodires, which are common in the northern hemisphere. Cryptodires withdraw their heads straight back into the shell by flexing their necks in a vertical "S" shape.

Even before the Pleurodires there was a sub-order of turtles called Proganochelys. They are now extinct. They were unable to withdraw their head and neck.

Australian turtles belong to the family Chelidae. They are all aquatic turtles and may be split into two groups: long-necked and short-necked turtles. The three species most commonly sold as pets are from each of these two groups.

Australian semi-aquatic freshwater turtles all have webbed feet and tiny tails. Both of the short-necked turtle species mentioned below have five claws on the front feet and four on the hind feet. Long-necked turtles have four claws on all feet.

The Saw-Shelled Snapper

The scientific name of one short-necked Australian species is *Myuchelys latisternum*. (It used to be *Elseya latisternum.*) These are commonly known as Saw-Shelled Snappers, Serrated Snapping Turtles or Common Sawshell Turtles. This species originates from northern New South Wales and Queensland, ranging as far north as the Cape York Peninsula.

Myuchelys latisternum babies have a carapace which rises to a ridge along the middle, from head to tail. The outside edges of the carapace are serrated. The carapace is made up of a number of 'shields', the colour of which ranges from grey to brown with shades of very dark brown.

The skin is brown except for the underside of the neck, which is creamy-coloured. As the babies grow older the serrations disappear from all except the rear edge of the carapace.

Under the lower jaw of *Myuchelys latisternum* are two small barbels of projecting skin, which look almost like a little beard. The neck is covered with nodules of scaly skin (not always evident in babyhood) and the skin on top of the head becomes coloured to match the carapace, so that when the turtle has withdrawn its head, the vulnerable top of the head looks like part of the hard shell.

Young Saw-shelled Snappers (Myuchelys latisternum). Note the ridge along the carapace.

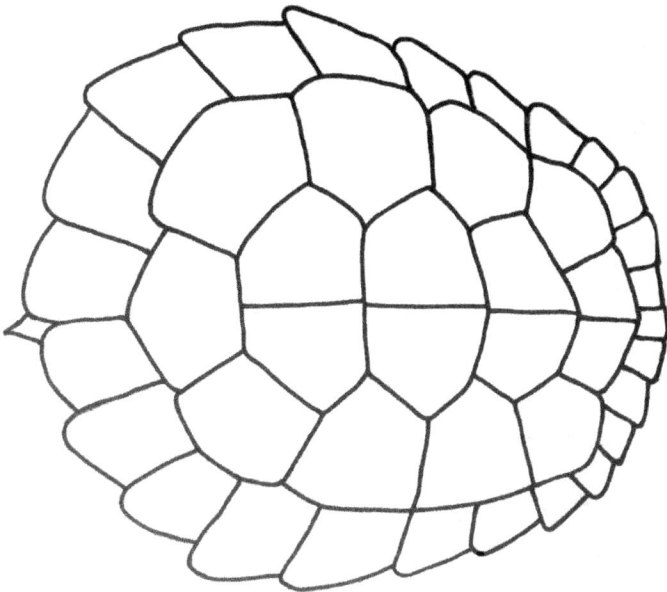

Carapace of Saw-shelled Snapper

The Murray River Turtle

This short-necked Australian turtle, *Emydura macquarii*, is also sold as a pet. It is sometimes called the Murray Short-Necked Turtle.

The Murray River Turtle has quite a vast habitat. It lives in many of the river systems of the eastern half of Australia. These turtles are most numerous in the Murray River Basin and along the major tributaries of the Murray, but they also live on the banks of many rivers along the coast of New South Wales Coast. In addition, they make their homes by coastal Queensland rivers, in Queensland's Cooper Creek drainage basin and on Fraser Island.

Murray short-necked turtle hatchlings look similar to baby saw-shelled snappers, with a circular, serrated carapace rising to a ridge from head to tail. As the turtle matures the serrations disappear to form a smooth edge all around the carapace. The ridge flattens and the carapace becomes more oval in shape. The shells of babies are usually bone coloured or a light yellow.

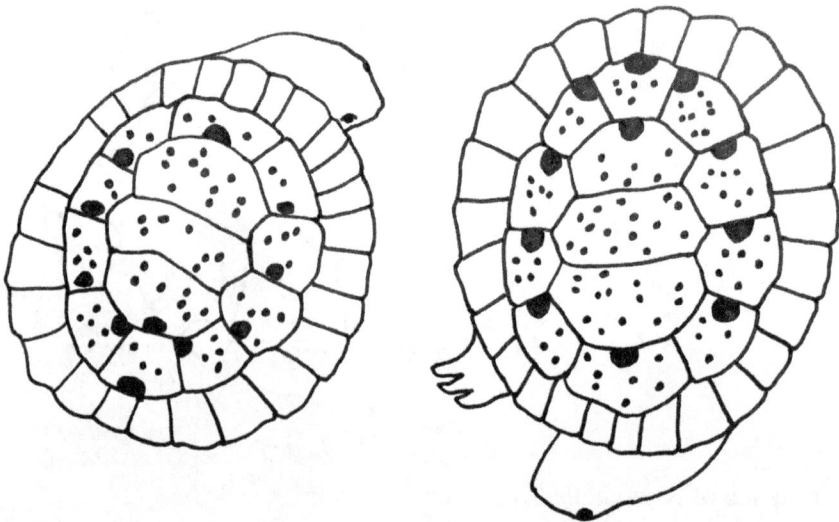

Baby Murray River Turtles (Emydura macquarii)

Eastern Snake-necked Turtle

The long-necked species sometimes kept in captivity is *Chelodina longicollis,* or Eastern Snake-necked Turtle. ("Longicollis means "long-necked".) Other names for this species include Common Long-Necked Turtle and Common Snake-Necked Turtle. This species lives throughout south eastern Australia. Their habitat ranges from west of Adelaide in South Australia, eastwards throughout Victoria and New South Wales, and northwards to the Fitzroy River of Queensland. They frequently make their homes in or near the Murray River.

The carapace of *Chelodina longicollis* babies is gently curved, almost flat. There is no hump or ridge. It is dark brown to black in colour. The skin is black. There are vivid orange markings on the underside of the edges of the carapace on either side of the head and along the edges of the plastron, which disappear as the youngsters grow older. These markings are most attractive, but they fade as the turtle grows to adulthood.

The head and neck of long-necked turtles are about as long as or longer than the carapace when they are babies. But when they are mature the average length of head and neck would be about 70 per cent of the carapace length.

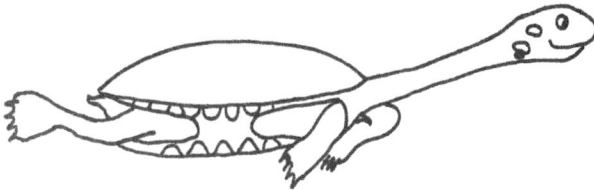

Baby Eastern Snake-Necked Turtle (Chelodina longicollis)

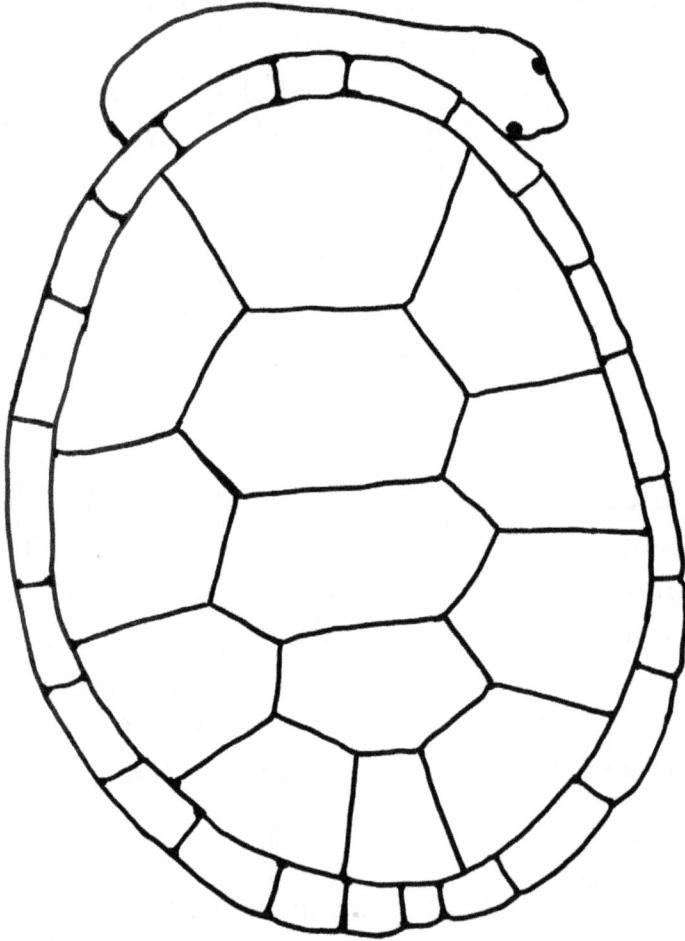

Adult Eastern Snake-Necked Turtle as seen from the top

Living Quarters

The care of Australian turtles may be divided into three categories —

(i) Care of adult turtles which are suitable to outside conditions in temperate (cool) climates such as exist in Victoria, South Australia and southern New South Wales.[1]

Group (i) includes the Eastern Snake-necked Turtle (*Chelodina longicollis*), the Broad-shelled Turtle (*Chelodina expansa*) and the Murray River Turtle (*Emydura macquarii*). In cool climates, only adults of these species may be kept outdoors all year round.

(ii) Care of adult turtles suitable only to outside conditions in tropical or sub-tropical climates, such as exist in Queensland, the Northern Territory and northern Western Australia.

(iii) Care of babies, or turtles of less than 75mm (3 inches) carapace length.

---

1   *Note: Tasmania has no native freshwater turtles. Any freshwater turtle seen in the wild in Tasmania is an invasive species.*

In cool climates, tropical or sub-tropical species of Australian turtles such as the Saw-Shelled Snapper, which are suited only to warmer conditions, must be kept in heated indoor tanks as described below.

For best survival rates, all baby turtles of all species should also be kept in heated indoor tanks, even if they are native to cool climates.

In temperate (cool) climates, tropical turtles can be kept outside during summer. In tropical climates, adults of all species may be kept outside all year.

Indoor Living Quarters

IMPORTANT NOTE
ABOUT INDOOR TANKS:

All turtles become stressed and depressed if kept all the time in indoor tanks.

Indoor tanks are completely different from their natural environment. By confining these wild creatures in artificial quarters you are denying their basic urge to roam and thereby creating a stressful situation for the animal.

Keeping adult turtles permanently in indoor tanks is NOT recommended.

It is VITAL to build a ramp and island in the tank, so that turtles can climb right out of the water and become completely dry if they need to. Their shells need to regularly dry out in natural sunlight or artificial UV light. Otherwise they can develop a fungus called 'shell rot' which, if left untreated, leads to a slow and painful death.

If you are going to confine turtles in tanks, only do it temporarily. You will need to provide these things in order to have a reasonably healthy pet:

- A large, strong glass tank
- Gravel
- Rocks
- Salt (iodised)
- Calcium
- Neutralizer block
- Filter
- Pump
- Filter wool
- Activated carbon
- Plants
- Heater
- Metal screen tank cover
- Aquarium lights

Preparing an Indoor Tank

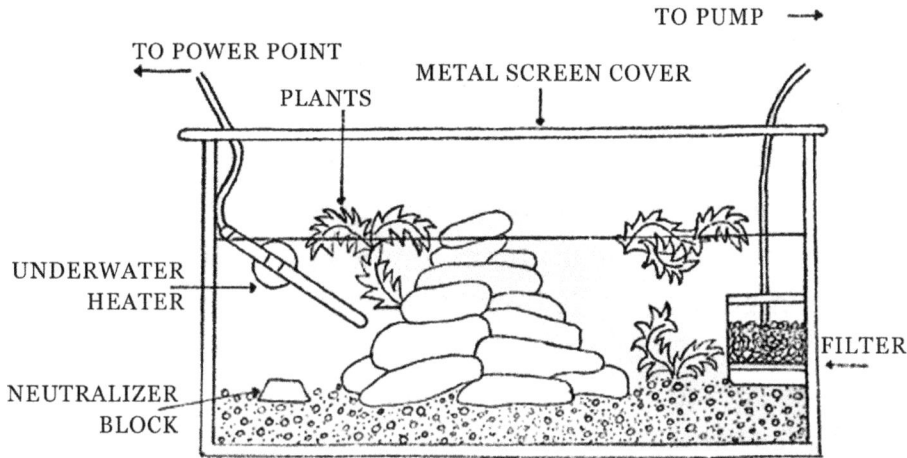

Indoor tanks : The Gravel
Once you have acquired a large tank of the right size for free-swimming, you must first put a layer of fairly fine, clean gravel on the bottom.

The gravel must be fairly fine and not coarse, as turtles like to dig in it, and coarse grains would be too heavy.

Gravel bought straight from a shop or gathered from a river-bed must be thoroughly washed in a bucket, by swishing it around in water then tipping out the dirty water. Repeat until the water runs clean.

Indoor tanks : The Rock Pile

One essential for a turtle tank is at least one good pile of rocks.

The turtles need a pile of rocks or stones which rises ABOVE water level because:

(a) they need support when surfacing for air,

(b) they like to climb out of the water occasionally,

(c) their instinct leads them to wedge themselves between rocks at night and while they are sleeping, for protection.

Rocks should be well balanced so that they do not fall on the turtles, and they should be smooth, preferably water-worn, from a stream or river so that they do not damage the plastrons.

Using silicone sealer, the special glue used in making aquariums, it is possible for the turtle keeper to build a custom-made rock pile for the aquarium, with caves and crevices for the pets to hide in. Otherwise you must rebuild the rock pile each time after you have cleaned the tank. Again, make sure a more or less horizontal or slightly-sloping surface of rock rises above the water level in the tank so that the turtle can climb right out of the water.

Half of a clean earthenware or terracotta plant-pot makes a good turtle-cave. Break the pot vertically, down each side. Clean, smooth, mortar-free bricks can also be used. Never put mortar into the tank, as harmful chemicals can leach into the water.

Turtles cannot swim backwards. Ensure that there are no places in the tank where they might get stuck underwater and drown. If ever a turtle does appear to be drowned it may be revived by holding it so that any water runs out of the lungs, keeping it warm and giving artificial respiration. This has been known to work successfully.

Indoor Tanks: Water Level

Water level should not reach the top of the aquarium in case the turtles climb out and try to escape while the cover is off. Also, it is hard to build a rock pile this high. In a 30 cm (12 inch) high tank, for example, the water should fill over two thirds of the tank.

Indoor Tanks: Water Conditioning

Salt: Before the animals are placed in their new home their water must be conditioned. In the wild, their environment is often a little brackish, so add about one rounded teaspoonful of common IODISED table salt per 23 litres (5 gallons) of fresh water.

Calcium: It is very important that calcium must also be added to the water. Turtles need plenty of calcium for their shells, which are an essential part of their skeleton. Powdered calcium may be obtained from pet shops (it, is often given to puppies and kittens) or from pharmacists. Brand names include "D.C.P.". Add one heaped tablespoon of calcium powder to a 46 litre (ten gallon) tank. Don't be alarmed when the water goes cloudy white - the calcium does take several hours to settle, and the cloudiness is harmless to the turtles. Do not forget to add the calcium and salt every time you change the water when cleaning the tank.

Conditioner: If you live in an area where the tap water is fairly heavily chlorinated (such as parts of South Australia), then it is best to let the water stand in a non-reactive container (eg plastic, ceramic or glass) for a few days before putting living creatures into it. Alternatively, you can use water conditioner products purchased from aquarium shops.

Indoor Tanks: Water pH and Neutralizer Blocks

There must always be a "neutralizer block" in the tank with the turtles. Their excretions make the water acidic. The neutralizer block counteracts this, and at the same time slowly releases valuable minerals and some calcium into the water. Neutralizer blocks are available from aquarium supply shops.

Buy a water pH testing kit and test the tank water once a week. "Water pH refers to how acidic or alkaline the water is. Ideally turtles' tank water should be kept at a pH of between 7 and 8.4 which is considered neutral to slightly alkaline."

If the water is too acidic or alkaline, ask your aquarium supply shop for a safe product you can add, to return the water to a healthy pH level.

INDOOR TANKS:
RESPITE FROM THE TANK

Because turtles need sunlight to keep them healthy, take the animals out of the tank and put them outdoors for a few hours whenever convenient, in a shallow container with rocks and water and especially some shade where they can escape if they get too hot.

Make sure they can't climb out, as they will always try to return to their natural wild homes.

Indoor Tanks: The Filter

Turtles make their water dirty quite quickly, and as it is a time-consuming job to clean an aquarium, it is wise to buy a filter and pump to keep the tank clean. Plastic filters are very cheap; pumps a little more expensive, but worth it, as you will find as your turtles grow and eat more.

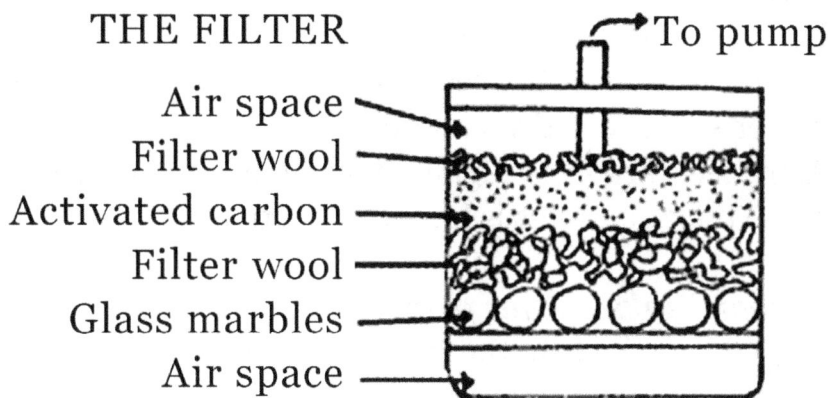

THE FILTER — To pump

Air space
Filter wool
Activated carbon
Filter wool
Glass marbles
Air space

Do not buy an under-gravel filter for turtle tanks.

If necessary, put marbles in the bottom of the filter to stop it from floating. On top of the marbles goes filter wool, then a layer of activated carbon (available from aquarium shops), then a thinner layer of filter wool to keep the charcoal from floating, then a small space for dirt to collect.

You can now buy filters that have "stabilizer platforms" attached to their bases. As well as making them heavier and better-balanced, this platform can be covered with gravel to anchor the filter to the floor of the tank. This eliminates the need for using marbles to weigh down the filter.

As the turtles grow, it may be necessary to buy a motorised filter for their tank. These efficient little machines, if cleaned once a week, do such a good job that you may only have to clean out the tank 3 or 4 times a year.

Because the water level in a turtle tank is low, you will require a fully immersible filter if you buy one. Motorised filters clean all the water in the tank every few hours, and circulate the water as well.

It is important to ensure that there is no uneaten food left rotting in the tank as this leads to the presence of bacteria in the water - an unhealthy situation.

Activated carbon is highly efficient in removing odours from the tanks, and this is important with turtles.

Indoor Tanks: Water plants

Keep water plants in the tanks with the turtles because
- now and then they like to chew on them,
- they seem to help keep odours down, and
- turtles like the shelter they afford.
-

Turtles seem to enjoy nibbling a plant called "Water Sprite". It is impossible to keep the plants stuck in the gravel as the turtles always scratch around and dig them up. Leave the plants floating in the water; they do well. If you want to decorate your tank use some plastic plants; they can look quite effective.

Indoor Tanks: The Heater:

Heat the water in the tank by using a special heater from aquarium shops, which sticks by suction to the inside of the tank and is thermostatically controlled. If you have a heater you must have a water thermometer too as the heaters probably won't show the temperature. Heaters are important for another reason as you will see when you read about hibernation.

Keep the water temperature between 75° F and 85° F (24° C-29° C). Heated water makes turtles more lively; and if they are not warm enough the tropical ones can go into premature hibernation, which can cause problems.

You may wish to turn the heater off during summer. In the latitude of Melbourne, Victoria, this should not be done before January 1st, when the warmer weather is presumably 'here to stay'. In more northerly latitudes, heaters may be turned off in December. Turn them on again in March.

Safety precautions: treat the heater with absolute care so as not to break it. Water and electricity can be a dangerous mix. ALWAYS make sure the heater is switched off and cooled down before removing it from the water or placing it back in the water. NEVER plunge a hot heater into the water.

Indoor Tanks: Covers

Cover your tank, to keep out dust and household pets. A cover also keeps in warm moist air, which turtles like.

"Please note that UVB light doesn't penetrate glass or plastic, so don't use a glass or plastic cover on your turtle habitat. Use a metal screen cover instead. You shouldn't use no cover at all because that can be dangerous. Lamps occasionally explode when they get splashed by water, and the glass can injure your turtle."[2]

Two corners of the cover must be cut off to allow filter tubes and heater cords into the tank. Don't leave sharp edges which may cut your hands.

---

2 *RJM Web Design, www.myturtlecam.com/lighting Retrieved 5th May 2016*

Indoor tanks: Lights
Light rays:

To stay alive, aquatic and semi-aquatic turtles need to spend several hours a day basking in natural sunlight or lights that closely mimic natural sunlight.

Turtles kept indoors need special lights, which should be pointed at their above-the-water basking area. They need lights that, like natural sunlight, contain UVA (Ultraviolet A) light, and UVB (Ultraviolet B) light. Ordinary electric light bulbs do not produce these rays.

Turtles also need to receive a little warmth from their lights. The sun would provide this warmth if they were living in the wild.

"The warmth helps raise their body temperatures to help fight germs; the sun-spectrum and UVA rays in natural sunlight are important to activity, mood and feeding; and they rely on UVB light rays to be able to produce Vitamin D3 in their skin. Vitamin D3 is essential for proper calcium metabolism in turtles.

The artificial light must not be filtered through plastic or glass because UVB rays cannot pass through these materials.

"Without the proper lighting, your turtles will get very sick and will eventually die. It's really that important."[3]

Sun-like rays "... can be provided by artificial 'UV-lights' (available from aquarium and reptile supply shops), however, there is NO substitute for natural, unfiltered sunlight."[4]

---

3   *RJM Web Design*
4   *David Vella*

Light distance:

'When using any UVB lamp, it's important to place it at the right distance so that the turtle will get the right amount of UVB light. Usually this is about 30 cm (12 inches) for a UVB 2.5 lamp and about 45 cm (18 inches) for a UVB 5 lamp. The companies that make these lamps usually provide detailed information on their websites."[5]

Never place the light too close, or you could burn the turtle. Never position the light too far away, or the turtle will not receive an adequate dose of the beneficial UVB and UVA rays.

"The effective UVB emission lifespan of these lights is **usually around 3-6 months**, so the light will need to be **replaced** with a new one at least every 6 months."[6]

Light timing:

"Turtles also need to have 'days' and 'nights', just like people do. Without day and night cycles, your turtle's sleep habits will be disrupted, causing stress and possibly reducing your turtle's immunity and overall health. As your turtle's keeper, you are creating everything in their world — including days and nights! You use the lights to do this."[7]

"The recommended day and night cycles for most of the temperate turtle species is 12 hrs light and 12 hrs dark."

It's best to use a timer to automatically switch the lights on and off, so that you don't have to remember to do it each morning and evening.

---

5   *RJM Web Design*

6   *David Vella*

7   *RJM Web Design*

Light safety:

You should always secure the lights over your turtle tank in such a way that they can't fall into the water. If they do, you or your turtles could be electrocuted and die! So if you set them on the metal screen cover, also secure them in some way so they can't fall into the water if the cover is removed."[8]

Light — Regular access to natural sunlight:

"Captive turtles should be placed outdoors in natural in sunlight for 20-30 minute periods 2-3 times a week. When doing this, ensure the turtles are enclosed safely in an escape proof/predator proof container with access to some shallow water. The sunlight shining on the turtles should not pass through any glass or plastic, as this will filter out UVB rays."

Provide plenty of shade to which the turtles can retreat if it gets too hot.

---

8   *RJM Web Design*

Indoor Tanks: Cleaning

This job can take quite a while, but as long as you have the filter going every day and change the filter wool frequently you should not have to clean the tank more often than about every three months in winter and slightly more often in summer. If you over-feed your turtles your tank will get dirty more quickly.

When you are about to clean the tanks it is wise to put on waterproof clothing and rubber boots. Rubber gloves also come in handy. Here is the procedure to follow:

(1) Carefully remove turtles from tank and put them in a high-walled plastic container in a warm place, with a little lukewarm water to keep them moist. When picking up turtles, always place part of your hand gently against their head. They then think they cannot go forward and won't struggle so much.

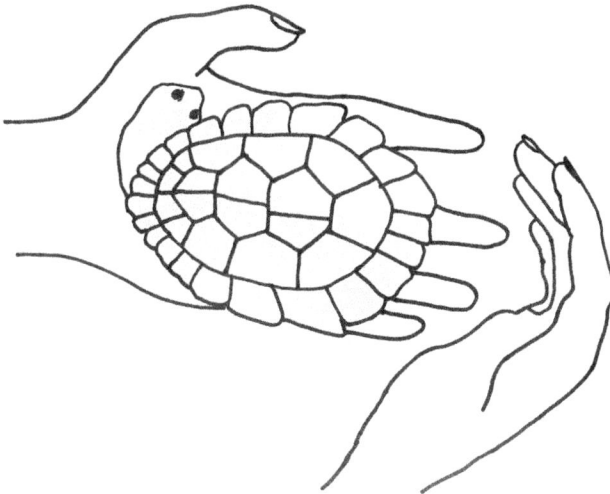

Use both hands to pick up each one; it's safer for the turtle.

(2) Switch off the heater, leave it in the water to cool for at least 5 minutes. If heaters are switched on and are out of the water, the glass will break or they will burn out.

(3) Put the water plants into a separate container of luke-warm water.

(4) Drain dirty water from tank using a syphon or scooping it out with containers.

(5) Remove heater, filter, rocks, thermometer and neutralizer block.

(6) When you have removed as much water as possible, carry the tank with the gravel in it, outside. Get someone to help you, as tanks are heavy.

(7) Using the hose switched on full blast, wash the gravel by tipping the tank on an angle, swishing the gravel around in the water with the hose and gloved hands, and tipping out the dirty water. Keep doing this until the water runs clear. Make sure all gravel is washed.

(8) Make sure tank sides are clean, by scrubbing them with a non-abrasive pad (available in supermarkets among the products for cleaning dirty dishes). Use a clean, new pad. Do not use any cleaning chemicals. Then return the tank with the clean gravel to its position indoors.

(9) Scrub rocks with warm water - NEVER use detergent or cleaners on anything which is to be put in the tank.

(10) Clean filter, heater and thermometer.

(11) Arrange rocks in a pile on the gravel. Make sure they provide a ramp by which turtles can climb out of the water when they need to get completely dry.

(12) Fill the tank to the desired level with clean fresh warm water; not too hot (test with a thermometer before replacing turtles). Temperature should be between 75° Fahrenheit and 85° Fahrenheit (24° Celsius to 29° Celsius).

(13) Replace heater, filter, neutralizer block, thermometer. Switch on heater.

(14) Condition the water with calcium and salt.

(15) You may put the turtles back into the tank now, then add the water plants.

(16) Put the cover on and replace the light.

Indoor Tanks: Vacuuming the Tank

Small amounts of waste matter or uneaten food may often accumulate at the bottom of the tank, particularly in between rocks. These may be removed without having to clean out the whole tank, by "vacuuming" the tank.

You will need a piece of clear plastic tubing, one to one and a half metres long (depending on the size of the tank to be cleaned), and one centimetre in diameter. This tubing is readily available in aquarium shops. You will also need a bucket and possibly something to stand the bucket on.

Completely submerge the plastic tubing in the tank water. It must fill with water. You can see whether there are any air bubbles in it; wiggle the tubing until they escape.

Leaving one end under water, put your thumb firmly over the other end, allowing no air into the tube. Take this closed end out of the water and lower it into the bucket, making sure that the open end is still in the water. The bucket must be at a lower level than the tank, so that when you release your thumb the tank water is sucked through the tube and down into the bucket.

The tank end of the tube will now suck in the dirt with the water. It is surprising how strong this vacuum can be. It can attract quite large pieces of gravel if you are not careful : these block the tubing and prevent the passage of dirty water. So keep the vacuum away from the gravel : stir up the dirt a little with your hand as you hold the tube, and it will float upwards to be vacuumed away into the bucket.

To stop vacuuming, merely lift the tank end of the tubing up and out of the water, and let the last of the water remaining in the tube run into the bucket.

Of course this process lowers the water level. Replace the dirty water you have taken out with clean water of the same temperature as the water in the tank, and conditioned with a little salt and calcium.

Indoor Tanks: Algae on Tank Sides.

If green, fuzzy algae begins to grow on the glass sides of the tank, it is easily removed by using a scouring pad made of plastic - not steel wool. Don't use the same one that's used for washing up! Keep a clean new one especially for the purpose. Make sure it's not contaminated with detergent.

NOTE: Algae is not harmful to turtles. However, its presence might indicate that it's time to clean the tank.

Outdoor Living Quarters For Adult Turtles

The following conditions should be met when providing an outdoor home for adult turtles:

(i) A Large Enclosure.

Turtles love to roam, so the enclosure should be large enough for them to be able to have a really good walk on land. They are also notoriously good at escaping, so close off any possible escape routes. If they escape in the suburbs they are likely to be run over by cars or mauled by dogs.

"An enclosure with walls about one metre high may be constructed from chicken wire sunk 15cm into the soil to prevent turtles burrowing out."[9]

In addition, the enclosure should be cat and dog proof, so you may need to build lower walls with a chicken-wire 'roof', securely fastened, if cats and dogs are a problem. The chicken-wire should have holes small enough to prevent turtles from poking their heads through. Half-inch holes are usually small enough.

For extra safety from animals, you could have a double fence, with a space between the fences. There should be shortish grass on the floor of the enclosure never a hard or rough surface.

Never underestimate a turtle's ability to climb. Despite their awkward appearance they can climb just about anything, and also lift bricks many times their own weight.

---

9 *'Tortoise Care' by S. Beattie, Fisheries & Wildlife Div., Victoria.*

(ii) A Pool.

In the wild, Australian turtles spend most of their time in water: they can only swallow food when underwater. Pools should be fairly deep, and the turtle MUST be able to climb in and out at will. The pool must contain piles of rocks for basking. Water in the pool should be kept relatively clean and MUST be CONDITIONED with salt and calcium (see "Indoor Tanks: Water Conditioning" on page 13). Otherwise it can become too acidic. Just as importantly, calcium is essential for turtles' shells.

Fibreglass pools are quite expensive. There are alternatives: a large plastic baby's bath sunk into the ground, or a concrete pool. Before putting any living creature into a new concrete pool, the pool should have been filled with water for three weeks, then drained and refilled, or coated with special preparations which are available specifically for the purpose of sealing concrete pools and preventing the escape of harmful chemicals.

Pool design:

When designing a pond, provision should be made to give a turtle sufficient area and depth of water in which to swim comfortably, and if it chooses, sufficient mud at the bottom in which it can hibernate in Winter.

It also needs a place where it can come ashore to bask in the sun during the warmer months : but the pond must be shaded by trees or shrubs. Direct sunlight on the water during the height of summer can raise water temperatures in a shallow pool so high that the turtle could die.

"For both convenience and beauty, a pond in an 'L' or boomerang shape is possibly the most convenient design. Its size will depend on the number of turtles to be kept and their dimensions. Preferably, it should be from six to eight feet long (182cm - 243cm) and not less than three feet (90cm) wide. If easy construction is desired, the best design has straight sides with the bottom formed into a ramp at one or both ends. This then enables the turtles in the pond to come ashore when they need to.

"At its deepest, the pond should be at least fifteen inches (38cm) or more, and should maintain this depth for at least half its length. Turtles eat and damage water lilies and other plants, but if sufficient are provided, some will survive and considerably beautify the pond in summer. Forms of duck-weed and other surface floating water plants can also be obtained on field trips. These provide some dietary needs and may beautify the pond.

"Because concrete will damage a turtle's plastron, the bottom should be ridged (slightly) and covered with several inches of fine sand or clay. Rainwater is usually sufficient to maintain the water level, but in dry seasons the pond may need to be 'topped' with a, hose when the water level drops.

"Some form of overflow pipe, leading to a drain, should be provided to prevent flooding of the surrounds during a wet season. Methods of constructing such a pond are available in "how to do it" or "home handymen books (e.g. Sunset Books).

"Ideally a pond is best constructed alongside a small tree or large shrub. Depending on the space available, the pond surrounds should be no less than three feet wide, and wider if possible. If no shrubs are available, one should be planted in such a position that it casts a fair shadow across the water during the hours between 12 noon and 4 p.m. in summer."[10]

(iii) Provision for hibernation.

Turtles may prefer to hibernate in the mud at the bottom of their pool, but in case they choose otherwise, a thick pile of dry grass clippings and leaves should be provided on land, with access to the pool, and in a sheltered place. Note that dry eucalyptus leaves are preferable, because they take a long time to rot and break down. Some owners have constructed little 'hibernation huts' for their pets. An overhanging bush may be sufficient. Let the turtle choose when and where it will hibernate.

Never disturb a turtle during hibernation, as the shock may kill it.

As John Goode writes, ". . . the initial effort expended is more than that required by most pets. Once this is provided, [turtles]. . . most probably, have a life span considerably greater than other domestic animals." [11]

---

10 *Goode, p.124-125*
11 *Goode, p.124*

Feeding Your Turtles

Some aquarium and pet shop staff in Australia have been known to tell customers to feed turtles on dandelion and lettuce. This just shows how little is known about them: no wonder so many of them die.

Do NOT feed your pets on commercially prepared 'turtle food' as this leads to malnutrition and eventual death. Australian turtles cannot live on "turtle food".

Turtles need variety, and the best foods to give them are as follows: frozen shrimp (Daphnia), frozen beef-heart, tubifex worms, and Tetra Min Staple Tablet Food, which can all be bought at aquarium shops. The first two items can be stored in the freezer; tubifex worms should be kept in water under a dripping tap. As mentioned before, turtles enjoy a little vegetable matter, and occasionally they chew on water plants.

Other foods which are an excellent part of their diet are finely chopped fresh liver, cat-food or dog-food, tadpoles, mosquito larvae, finely chopped lean muscle meat, ox heart, and especially chopped earthworms. Tropical turtles may be fed mainly on the shrimp, small amounts of Whiskas cat food (commercial pet food contains a lot of vitamins), minced meat no preservatives or fat) and Tetra Min tablets. These other foods are given as a special treat when available. Try to give them as often as possible. The main thing is plenty of variety, no preservatives and no fat.

Many adult turtles are very fussy about their food. If you find that your turtle will only eat red mincemeat (which must be fat-free) and refuses other foods, don't worry. Turtles who are kept outdoors find food in their pool-water, such as mosquito larvae, insects, and possibly tadpoles and worms. If you try giving earthworms to your turtle, you might have to chop them up, as they may be too big to swallow whole.

Adult turtles have been known to catch unwary birds bathing in their pool. They grab their prey by the leg and pull it underwater.

The construction of turtles' necks makes it easier for them to take from overhead rather than from below. Australian turtles can only eat when they are submerged in water. On land they cannot swallow.

Turtles should be fed once a day. They may often appear to be greedy, but they merely eat as much as they require. A turtle with a four-inch long carapace should have a daily ration lump about the size of a pile of ten twenty-cent coins. Babies should have about one third of this. Frozen food can be put into the water in lumps - it will soon thaw.

As a basic, staple diet, I recommend the frozen shrimp, which can be given almost every day,

One good method of supplying food is to buy two or three pounds of fresh ox-heart, cut off every bit of fat and gristle, mince it very finely, then wrap small amounts in plastic, and freeze the lot. Each packet can then be defrosted the day before it is used.

Turtles need Vitamin D to nourish their shells. Vitamin D and direct sunlight are closely associated. Sunlight filtered through glass loses most of its Vitamin D-giving rays, so whenever possible, give your pets some hours in natural, outdoor sunlight, with shade available to them.

Cod liver oil is a good vitamin supplement for turtles. I give it to them with an eye-dropper when they are outside. Don't give it to them when they are in their tank, unless you want very oily water. If they won't open their mouths for the oil (which they don't), just drop some on the outside of their mouths and hope that it seeps in. Or you can give the oil to them on their food.

Cod liver oil can be toxic in high doses and too much can poison the turtle.

Hibernation

In their natural state, when cold weather comes turtles go into hibernation. This occurs from about mid-autumn to early winter. They live off fat reserves stored up in the summer. Buried deep in the earth or in mud under water, where the temperature ranges between 35° F (2° C) to 50° F (10° C), the life processes of the animal slow down to practically zero. In spring they wake up.

Before hibernation they must fast to make sure their stomachs are empty, and they must have ejected all faeces.

If the temperature during hibernation is too high, the metabolism speeds up, the turtle uses up more stored food and it will starve to death before it wakes up.

Many pet turtles die when the time comes for hibernation, because owners are ignorant of the conditions needed. I have also heard several times of disastrous cases when owners have thought that their hibernating turtle was dead, and buried it or thrown it away. To avoid the dangers and inconveniences of hibernation it is preferable to keep your tropical turtles and young turtles indoors and active throughout winter.

All you do is to provide artificial heat and light, which you should already have. The light lengthens the days for the turtles, the heat keeps their body processes at their usual level. It is perfectly all right for turtles to miss out hibernation for a year or two, and it seems they can be enjoyed by their owners all year round.

Adult turtles kept outdoors should be allowed to hibernate; the owner must simply provide the right hibernating facilities as described, and allow nature to take its course.

Only healthy turtles in good condition should hibernate and turtles should be about 3 years old (shell length 5-8 cm).

Do not feed the turtles within a month of hibernation time (ie. stop feeding at end of March for the Canberra area) as food may not be digested and will rot in the stomach. Do not wake the turtle while it is hibernating. It does not need to eat, as it will use up fat reserves.

The outdoor enclosure must be made suitable for hibernation. You need to provide plenty of non-toxic ground cover plants, other plants and grasses, soil that is suitably soft for digging, some deep piles of dry leaves and a pond at least 60cm deep, containing water-plants and a good thick layer of mud. There should be a shallow ramp access to the pond so that turtles can easily climb in and out.

The turtle will choose where to hibernate. Do not force it.

The turtle must not be exposed to frost and must be able to stay dry if it wishes. Most turtles will choose to hibernate in the water. Make sure no predators (eg. rats, dogs, cats) can get to the turtle during hibernation.[12]

---

12 *Eastern Long Neck Turtle. Scott Thomson. © 2003 World Chelonian Trust.*

Ailments

Sickness

If you keep the water clean, feed your turtles on a variety of the proper foods and provide sunlight rays and calcium, they should never encounter sickness. Sickness is often caused by malnutrition.

Sores

Sores on the plastron are caused by rough surfaces. Use smooth rocks in pools and enclosures, cover concrete with sand. Treat sores with a common antiseptic - not mercurochrome, however. Remove the turtle from the water, dry the shell gently with a soft, clean cloth and use clean gauze or cotton to apply the antiseptic.

Cuts and Bites

Treat cuts and bites the same way. Allow the antiseptic to sink in for several hours before replacing the turtle in the water.

Swollen Eyes

Swollen eyes are a result of poor diet and living conditions. Check diet, and bathe turtle frequently in lukewarm water in which a little iodised table salt is dissolved. This salt bath is also the remedy for infections of the eye and fungus infections of the skin. Make sure the turtle can regularly bask in natural sunlight.

Fungus

Another excellent fungus combatant is a preparation called "ANTIMALADIN" available in aquarium shops. IODINE may also be successful. Anti-fungal preparations that are safe for fish are also safe for turtles.

Warm, unfiltered sunlight is the best medicine for turtles, providing shade is always available. Baby turtles can easily be infected with fungus if they do not receive sunlight or a sunlight substitute. YOU NEED TO AVOID FUNGAL INFECTIONS AT ALL COSTS, as they can be difficult to treat and can be fatal.

Respiratory infection

Respiratory infection may develop due to inadequate heating of living quarters. Look for a discharge from the nose and mouth in the form of bubbles or liquid, loss of appetite, a "wheezing" breath and drooping head. Gradually increase the temperature of the living quarters and treat the turtle.

" ..... with antibiotics such as sulphadiazine, sulphamezathine or aureomycin. In liquid form, an approximate dose of 30mg a day for three to four days should be administered to a turtle with a carapace length of 22.5cm and correspondingly smaller doses for smaller turtles. This allows for spillage which may occur, but if a full dose is taken, treatment for the following day should be omitted." [13]

---

13 Cann, pp.41-42

Following this treatment the turtle should be kept apart from other turtles until it is recovered.

You will need to ask a vet for these medications.

Soft Shell Disease

Soft shell can be a fatal condition and is due to lack of calcium and sunlight. If it does eventuate, rectify the conditions immediately and add an antibiotic to the water to prevent fungus infection.

Parasites

If you see red or white worms wriggling at the bottom of the turtle tank, it means that your pets are infested with the parasites. Treatment for worms is as follows: "Adult turtles are given 1 gram santonin with 1 gram calomel, and, the following day, a full eye-dropper of castor oil. Smaller turtles are given a smaller amount." (Cann, p.45). Clean tank thoroughly after treatment. A vet can give alternative treatment for turtles with worms.

David Vella's Veterinary and Health Notes

- Have any new turtle examined by a reptile vet. Ask for parasite checks and general blood screens.

- It is essential that you quarantine any newly introduced turtle. Don't risk introducing disease or parasites. Speak to your reptile vet for details on proper quarantine procedures.

- It is recommended that you have your turtles vet-checked annually.

- Always wash your hands after handling any reptile and between handling of different reptiles.

- Turtles kept outdoors or those that are intending to go under hibernation need to be closely monitored for health and parasites. It is a good idea to regularly weigh and record the body weight of your turtles.

- Turtles can be transported in sealed (provide air holes) flat bottom containers on soft and moist towels. Ensure that they cannot escape or overheat.

- Never transport turtles in water.

- When you take your turtles to the vet for a check-up, consider bringing bringing a sample of their pond water or tank water. Use a clean jar or obtain a specimen container from the vet beforehand.[14]

---

14 *Vella, David. Care Of Australian Freshwater Turtles In Captivity.*
2013

Behaviour And Intelligence

Although to humans, turtles appear to be stupid in some ways, this is because we do not always understand their behaviour, and because we expect them to behave as we would ourselves behave in a given situation.

This 'anthropomorphism' is the cause of our underestimation of the intelligence levels of many other animals. Their code is different from ours - they would probably consider humans to be quite stupid! Animals sense many things that we cannot: their bodies are finely tuned to an enormous range of smells, tastes, vibrations, wavelengths and even electromagnetic fields. Their reasoning, obviously, is also different from ours.

We have yet to devise a suitable intelligence test for turtles.

Richard Haas [reference 4] says that turtles "... can learn to discriminate between white and black, between vertical and horizontal lines, between lines of varying width and between various colours. Some have the rudiments of a kind of social organisation shown by the 'pecking order' established in a captive group of turtles.

"[Turtles] of the same species become recognisable as individuals to their owners because of the 'personality' of each.

Possession of a personality indicates some degree of intelligence."

Furthermore, turtles "... possess a finely-tuned ability to detect vibrations coming to them through the water or ground and an excellent sense of touch. The sense of taste and smell are believed to be well developed."

Turtles of the Chelidae have exceptionally keen sight out of water, and very good hearing. In fact, their five senses all either match, or are superior to, our own.

They can be trained, with patience and consistency, to eat out of your hand and to come when you call or whistle.

As mentioned before, they are social animals, and prefer the company of other turtles. By themselves they get lonely.

However as they grow older watch for aggressive ones that may intimidate others by biting and pushing. These should be separated from the rest until they learn to coexist peaceably.

Turtles instinctively love to hide, wedging themselves amongst rocks for protection if they are frightened, or half-burying themselves in gravel. They are very shy creatures.

Some adult turtles love to be stroked gently on the underside of their necks.

Here is an interesting anecdote gathered from a past Curator of Reptiles at the Melbourne Zoo, Mr Roy Dunn:

There was a pool containing turtles, around which crowds of visitors would gather daily. The turtles took no notice of their noisy audience. However when Mr Dunn would appear amongst the crowd, the turtles recognised his face as the man who provided the food, and they would swim furiously straight towards him. He did not have to make a sound — they knew him by sight.

Are they really slow-witted animals? How many of us would recognise the face of an individual turtle?

"Slow and steady won the race in 2006 when scientist Anna Wilkinson placed a tortoise and rat in the same maze.

"The reptile was better at navigating the maze to find food, making sure it didn't revisit the same area twice.

"When cognitive landmarks were removed for a second trial, the tortoise systematically visited each section of the maze to find food. The rat wasn't as methodical.

"Previous research hasn't shown tortoises to be so clever, though: Wilkinson suspects cold lab temperatures are to blame. Later research found that tortoises use gaze-following to learn from the behavior of other animals.

"They might be smarter than we thought." [15]

---

15 *Mental Floss, "16 Fun Facts About Tortoises" by Amanda Green. Retrieved 26th April 2016.*

How Old is the Turtle?

To tell how old a turtle is you must look at the shields of the carapace. A new perimeter is added to each shield each year as the turtle grows. This results in an annual 'growth ring'. Count the centre of the shield as year one. Each additional ring indicates a year of life.

This method of age determination is not always reliable as growth rings can wear away. Turtles periodically shed the outer covering of their shields as they grow; this also helps obliterate growth signs. As mentioned earlier, turtles can live for up to 200 years.

Dangers

There are several things which you should never do to any turtle - or to any tortoise or terrapin, either.

(i) Never paint on its shell, as its shell must breathe or it will die.

(ii) Never bore a hole in the shell. This is very cruel. Shell-boring is excruciatingly painful for the turtle. Their shells are sensitive. "Shells have nerve endings, so tortoises [and turtles] can feel every rub, pet, or scratch ..."[16]

(iii) Tying string or cord around one leg is just as bad. It inevitably cuts off the circulation and the poor turtle lives in constant pain as the leg dies, rots and finally drops off altogether. This has happened all too frequently. String around the neck would, of course, kill the turtle.

Turtles, like any other creature, love to be free to roam. To tie them to anything so they cannot escape is as cruel as it would be to do it to a human. Besides, there is no way you

---

16 *Mental Floss, "16 Fun Facts About Tortoises" by Amanda Green. Retrieved 26th April 2016.*

can tie up a turtle without harming it greatly. So to prevent escapes, a comfortable and secure enclosure is the answer.

(iv) Don't turn turtles over on their backs; this is psychologically stressful and traumatic for them.

(v) Do not allow little children to maul the animal. Turtles should not be over-handled.

(vi) When putting a new turtle in a tank with a turtle that is already living there, watch for as long as possible to ensure that there is no attack from the first turtle, who may have developed a territorial feeling for the tank. Never put baby turtles with adult turtles.

Look after your turtles; they may soon become members of a rare species. Due to constant, indiscriminate trapping and degradation of their habitat, their numbers are dwindling. It would be a great pity if such beautiful, unique and fascinating reptiles were to become rare or die out completely due to man's exploitation.

Reproduction and Sexual Differences

Male turtles have longer tails than females, and their plastrons are flat or curved inward. The females' plastrons are flat or curved outward. These characteristics cannot be seen when they are babies.

"Most of the sex determining features do not appear until the turtle is mature. Emydura macquarii male is immediately recognisable by the length and thickness of his tail. In Chelodina species it is almost impossible to determine the sex, and captive females often lay fertile eggs without apparent contact with a male. The female can retain sperm from a mating in the wild for a period of four years. Normally the female comes ashore to nest, but if unable to do so, will lay eggs in water where the embryos suffocate." (Beattie p.43-44)

Female turtles like to have a patch of soft soil where they can dig a hole in which to lay their eggs. Once the eggs are laid, the turtle covers them so carefully that the human eye cannot see where the hole has been. The eggs look like ping-pong balls. Baby turtles rarely hatch in captivity.

References

(1) Goode, John. *Freshwater Tortoises in Australia and New Guinea.* Lansdowne Press, Sydney, 1967.

(2) Roberts, Marvin F. *Turtles as Pets, a guide to the selection, care and breeding of land and water turtles.* T.F.H. Publications Inc., July 1980.

(3) Jocher, Willy. *Turtles for Home and Garden.* T.F.H. Publications, Inc., 1973

(4) Haas, Richard. *Enjoy your Turtle.* The Pet Library Ltd. New York, N.Y., 1967.

(5) Beattie, Susan. *Tortoise Care.* Arthur Rylah Institute for Environmental Research, Fisheries & Wildlife Division, Heidelberg, Victoria.

(6) Cogger, H.G. *Reptiles and Amphibians of Australia.* Reed, 1975

(7) Worrell, Eric. *Reptiles of Australia.* Angus & Robertson, 1963.

(8) Worrell, Eric. *Australian Snakes, Crocodiles, Tortoise, Turtles, Lizards.* Angus & Robertson. 1966

(9) Mincham, Hans. *Reptiles of Australia and New Zealand.* Rigby Instant Books, 1970

(10) Cann, John. *Tortoises of Australia.* Angus & Robertson, 1978.

(11) Vella, David. *Care Of Australian Freshwater Turtles In Captivity.* 2013

(12) © RJM Web Design. *Proper Lighting for your Turtle Habitat.* www.myturtlecam.com/lighting Retrieved 5th May 2016.

(13) Green, Amanda. *16 Fun Facts About Tortoises.* Mental Floss. Retrieved 26th April 2016.

www.ingramcontent.com/pod-product-compliance
Lightning Source LLC
Chambersburg PA
CBHW082009190326
41458CB00010B/3132